INQUIRY INTO SCIENCE FICTION

Inquiry into Science Fiction

By

BASIL DAVENPORT

LONGMANS, GREEN AND CO.

NEW YORK · LONDON · TORONTO

1955

LONGMANS, GREEN AND CO., INC.
55 FIFTH AVENUE, NEW YORK 3

LONGMANS, GREEN AND CO., LTD.
6 & 7 CLIFFORD STREET, LONDON W 1

LONGMANS, GREEN AND CO.
20 CRANFIELD ROAD, TORONTO 16

INQUIRY INTO SCIENCE FICTION

PUBLISHED SIMULTANEOUSLY IN THE DOMINION OF CANADA BY
LONGMANS, GREEN AND CO., TORONTO

FIRST EDITION

LIBRARY OF CONGRESS CATALOG CARD NUMBER 55-8306

Printed in the United States of America

Contents

INQUIRY INTO SCIENCE FICTION

CHAPTER ONE

What Is Science Fiction?

A YEAR OR TWO AGO *Time* magazine decided to do a piece on science fiction. Someone there knew that I was an addict and called to ask if *Time* might send one of their researchers to ask me some questions. She proved to be very pleasant, very intelligent, and very much bewildered.

"The first thing I tried to find," she said, "was a definition of science fiction. I thought that would be perfectly easy, but there doesn't seem to be one anywhere."

"Perfectly true," said I. "There are probably no two people who would agree on precisely the same definition." I went on to tell her what I could about the scope and development of science fiction, while she occasionally asked questions. When she was about to go she looked at her notes and said, "Oh, yes, one more thing: can you tell me what is a space-warp?"

"A space-warp," I said, "is a piece of jiggery-pokery for getting us Earthmen outside the solar system. You see, according to Einstein, it is impossible to travel faster than the speed of light; but even by traveling at the speed of light it would take four years to get even to Proxima Centauri, the nearest fixed star, and hundreds of years to go traveling around the Galaxy. So, to get around this, authors have invented various pieces of scientific double-talk. Some of them talk about 'putting the ship in hyperdrive,' which is supposed to be faster than the speed of light; and some of them talk about 'space-warps,' assuming that you can bend space itself so as to bring Proxima Centauri quite close in another dimension—" She was looking completely blank. I tried the classic analogy of putting two dots far apart on a sheet of paper, and rolling the paper into a cylinder to bring them together, but when I went on to the analogy of three-dimensional space in the fourth dimension I could see that I was losing her again, and I ended by recommending an elementary book on hyperdimensionality.

At about the same time, I visited a friend who collects the work of C. S. Forester. Before he settled to sea stories, Forester tried his hand at a variety of writing, not all of it published in the United States. These early Foresters included one science fiction novel, *The Peacemaker*, which was reprinted in one issue of a pulp magazine, *Famous Fantastic Mysteries*. I happened to spot this on a newsstand, and took it along. My friend had read *The Peacemaker* in an English edition, but he was fascinated by the large department of Letters to the Editor in the back of the magazine. "What is a 'prozine'?" he asked. "And what are 'fen'? It seems to be a plural."

Fen, I told him, is the plural of *fan*, on the analogy of

WHAT IS SCIENCE FICTION?

men from *man*; it is the form preferred by the really rabid science fiction fans, who are mostly teen-agers, and who have their own jargon. *Prozine* in this jargon means professional or commercially produced magazine (devoted, of course, to science fiction), as distinct from *fanzine*, a magazine, generally mimeographed, produced by the fans for circulation among themselves. Nobody knows how many there are of any of them, but the fans have an annual convention, held each year in a different city, where they meet to talk to each other and goggle at their favorite authors; the attendance at the last two conventions has been about fourteen hundred. There are also innumerable local societies that meet once a month or oftener. Among magazines of both types there is a high birth-rate and correspondingly high mortality, but at present there are over thirty commercial science fiction magazines, and probably sixty to seventy fanzines. There are three or four publishing houses devoted to science fiction, and nearly every one of the regular publishers has at least one piece of science fiction on his list. The established authors have contracts two and three deep, and original books cannot keep pace with the demand; there is such a flood of anthologies that it seems as if every story ever written must have been reprinted at least once. In addition, there are science-fiction movies, radio and television series, newspaper strips and comic books.

No wonder that people who, unlike myself, have not grown up with science fiction and so have not acquired painlessly a vocabulary of space-warps, parallel universes, and so forth, feel a bewildered curiosity about it. Is it, as one might suppose from some of the magazine covers, one more excuse for sadism, or another form of mental chewing gum, or a helpful stimulus to the imagination? What *is* science fiction, anyway?

There are no simple answers to these questions, especially to the last one. Quiller-Couch, when asked to define a ballad, said, "A ballad is *Sir Patrick Spens* and *The Lass of Lochroyan* and *The Bonnie Mill-Dams of Binnorie,* and things of that sort—" thus anticipating the modern science of semantics, which likes to abandon the old Aristotelian definition by means of class and distinction from other members of the class, and to substitute what it calls an operational definition: "If you will do so-and-so you will observe such-and-such. *That* is what it is." The best way to acquire an idea of science fiction is to look at its various types and see how they developed.

Enthusiasts have displayed their own erudition and have hoped to add respectability to science fiction by tracing it to various early sources, the earliest being Lucian of Samosata, who wrote narratives of two voyages to the moon in the second century A.D. So far as I know, no one has yet traced it to the story of Daedalus, the inventor of the first flying machine, which is even older. But the plain fact is that you do not get science fiction until you have applied science; that is, broadly, not until after the Industrial Revolution.

And modern science fiction, this unique phenomenon with a language of its own, is a good deal younger than that; it is usually dated from the foundation of *Amazing Stories*, the first magazine to be devoted exclusively to the genre, in 1926.

Before we come to this, however, we must look back and, passing over with due respect Lucian of Samosata, Cyrano de Bergerac, Jonathan Swift, and other candidates who have been put forward, must briefly salute Jules Verne and H. G. Wells—if only because most people outside the circle of fans, if asked to define science fiction, would probably say, "Science fiction is *The War of the Worlds* and *Twenty Thousand Leagues under the Sea* and things of that sort." Verne was a man who was fascinated by the expanding possibilities of nineteenth-century science. By no means all of his work, to which he gave the general title of *Voyages Extraordinaires*, is science fiction. One that is still among the most readable, *Around the World in Eighty Days*, is based on the idea that not by some future invention, but by the use of means already existing, it is possible to accomplish this amazing feat. And his extraordinary journeys include the straight adventure story of *Michael Strogoff, the Courier of the Czar*, in which the only bit of science is the principle, interesting if true, by which Michael's sight was preserved when the Tartars tried to blind him; as the red-hot blade approached his eyes, he thought of his aged mother, his eyes filled with tears, and "the drops volatilized

on the cornea!" But Verne's best work is that which shows
an invention just a little beyond the science of his time—the
airship, the submarine, the interplanetary passenger-carry-
ing projectile. Most of his inventions have already been
realized, and the rest seem well on the way; and one may
well wonder how many of the men now working on rocket
ships were first filled with the idea of flying to the moon
by the work of Jules Verne.

Verne prided himself that his work was as scientifically
accurate as he could make it. In many of his short stories
Wells did the same; "The Land Ironclads" is as remarkable
an anticipation of the tank as *Twenty Thousand Leagues
under the Sea* is of the submarine. But science fiction was
only one of Wells's many interests. Several of his science
fiction novels are really utopias; part of *The Time Machine*
and *The First Men in the Moon* are utopias in reverse,
showing the degeneration of a society. Since what he
wanted was to depict a strange society, Wells was willing to
make things easy for himself, and perhaps for his reader,
by postulating an invention that would get his characters
where he wanted them to go. It didn't matter whether or
not there was any real scientific basis for it, provided he
could make it sound possible. To get his hero into the future
he invented a time machine. Previous characters had of
course been moved about arbitrarily in time, like Mark
Twain's Connecticut Yankee, but it was left for Wells to
provide that plausible analogy, that persuasive question,

"Can an instantaneous cube exist?" To get his characters to the moon he invented a substance impervious to gravity, which he called cavourite—at which old Jules Verne snorted, "I sent my characters to the moon with gunpowder, a thing one may see every day. Where does M. Wells find his cavourite? Let him show it to me!" Verne's spacefliers would in fact have been squashed flat by the initial acceleration, but in imagining a projectile driven by explosive power he was on the right track, as far as actual development went. Wells, on the other hand, was a pioneer in the use of that sort of plausible double-talk by which modern writers evade impossibilities like that of traveling faster than light, and he gave us two or three of the most fertile ideas in science fiction—that of time-travel, that of interplanetary warfare, and, a minor one, that of the detached or almost detached brain—his Martians are all brain and hand, using machines for locomotion.

Verne's gunpowder-powered projectile to the moon was actually as impractical as Wells's cavourite. The first man who wrote stories "supplying the people of the future with technical inventions which are the logical developments of those currently in use or logically developed from currently accepted principles" (to borrow the words of Fletcher Pratt, one of the few writers who have given serious critical attention to science fiction) was a young man named Hugo Gernsback. In 1908 he founded *Modern Electrics,* the world's first radio magazine. In 1911 he contributed a fiction serial of his own authorship, *Ralph 124C41+*. Ralph, it may be explained, is a man living in the future, when names have been replaced by numbers; his own name is a pun, meaning "One to foresee for one"; the plus, if you are interested, is an honorific. Gernsback was unable to keep up the punning in the naming of his other characters, and the generally creaky nature of the device is perhaps a fair measure of the story's quality from a strictly literary point of view; but it is written by a man of science who knew what he was talking about, and its success with the readers of *Modern Electrics* was enormous. Gernsback continued to write, for later factual magazines such as *Science and Invention*, pieces of what he liked to call scientifiction, a portmanteau word which was taken up by the growing group of fans and often abbreviated to stf. Scientifiction proved a rather awkward mouthful and has been generally superseded by science fiction (with or without the hyphen), abbreviated to s-f or SF; but the abbreviation stf is still sometimes seen, and must be puzzling to those who do not know its origin.

In 1926 Gernsback founded *Amazing Stories*. This was the first magazine devoted exclusively to science fiction, and this event is the hegira from which modern science fiction is dated. The first issue was largely made up of reprints, from Wells, Verne, and Poe, but the magazine soon attracted a group of authors anxious to try the new medium. Other pulp magazines entered the field; by 1931 there were nine, though this fell off in the Depression. Gernsback himself was a stickler for scientific accuracy; many of us can remember the drawings that appeared under his auspices, giving hypothetical constructions of, for example, Martian man, with a huge chest to give space for the enormous lungs he would need to sustain life in the thin Martian atmosphere, and all the other physical attributes which would be developed by conditions as they are believed to be on Mars. Many of the stories in these early magazines, however, had no such scruples. There has always been a small and now decreasing fringe of romancers for whom the space-ship and the time-machine were merely means to get away from the here and now to a place where anything might happen (for the time-travelers, this place was usually Atlantis). Most of the stories had at least a flavor of science, but the science was often anything but scientific. A recurrent theme, for instance, was man versus insects of his own size, either because he had shrunk or they had grown. It makes no difference either way; because of the difference in the method of circulation, a mammal cannot be much smaller than a shrew-mouse or an insect much larger than a tarantula (which are about the same size). Stories were written about the other planets of our system which blithely ignored their peculiar gravities and atmospheres or lack of atmospheres.

As time went on, however, the authors and the fans engaged in a process of mutual education. The fans, or at least a portion of them, are a highly vocal group, and most of the early pulp magazines encouraged the writing of long letters to the editor. Fans who detected inaccuracies were quick to point them out, and before long everyone knew approximately what conditions were like on the planets of the solar system, or at least knew that on most of them you couldn't simply open the door of your space-ship and step out. As the solutions to the early problems appeared, they became common property. I doubt if it would be possible to establish who first used the phrase "a space-suit," any more than one could establish what ballad-writer first wrote the stanza

> Some whiles the page he went,
> Some other whiles he ran,
> And when he came to a river
> He bent his bow and swam,

which was used in so many ballads that wanted to get a character from here to there in a hurry; but for at least a literary generation, any writer putting characters ashore on a strange planet usually dresses them in space-suits, and any reader who has grown up with science fiction understands that they are wearing helmeted, sealed suits which supply them with oxygen and which usually have radios or some other means of communication. In the same way, the fact that even at the speed of light any journey outside the solar system would take a number of years was got over, either by the idea of putting a bend in space, so as to bring Alpha Centauri near us through fourth dimension—the "space-warp" which I tried to explain to the young lady from *Time*—or by conjuring up a "hyperdrive," which may be defined simply as something that *does* enable ships to travel faster than the speed of light, no matter what Einstein says.

Similarly, the paradoxes of time travel have been a fertile source both of fiction and of difficulties. What happens if you go back to a point in your own lifetime—do you encounter your own younger self? Or what happens if you kill your own grandfather before you were born? An enormous amount of ingenuity has been spent on treating these themes from every angle, and they have produced some good stories (as well as some mechanical ones); but they confronted authors with one great difficulty: a modern man sent into the past might normally be expected to produce some mark of his presence there; even if he did not change the course of history, one would expect to find some mention of his visit in the history books; but of course there is none. This knot was cut by the idea of parallel universes; and here I believe it is possible to give credit to the first story to use the idea—though in this vast and unpruned field it is always possible to be mistaken. It was, I believe, *Sidewise in Time*, by Murray Leinster, which first put forward the idea that there are a number of universes (or "space-time continua," which sounds more scientific and hence more plausible) existing side by side, in which various historic alternatives were different—in which, for example, England held the American colonies, or the Russians held California. Some writers have expanded this into the concept of an infinite number of universes, in some of which every conceivable alternative has occurred; others, more modest, have been content to let the other possible universes lie latent; that is, if your hero goes back and averts the fall of the Roman Empire, he will start a new branch at that point on the trunk of time, which will develop into a universe parallel to our own, but one in which the Roman Empire did not fall. By now, "parallel universe" is a phrase as well understood by the initiates as "space-suit" or "space-warp"; there are novels, and good ones, whose only claim to be scientific is that they take place in a world where, for instance, the Civil War was won by the Southern Confederacy.

The best of these parallel-universe stories are based upon extrapolation of some tendency in our own society. "Extrapolation" is a word that is almost as great a favorite in discussions of science fiction as "space-warp" is in science fiction itself; it may be defined as "plotting the curve." We take some tendency in our present society, toward sexual license, say, or the prolonging of the life span, or the dominance of women, imagine it enormously increased, and show the probable consequences.

This last may be described as science fiction based on the social instead of the exact sciences. Parallel universes, like remote planets, which are conceived merely as places different from the world we know, where the hero can go and have adventures, are obviously among the spots where science fiction merges into "fantasy fiction." A word must be said about this also, for its existence is one of the things that make the definition of science fiction so difficult. The folk-tale form of "ghost story" (another unsatisfactory term) must go back at least as far as Shakespeare's little Prince Mamillius, who started to tell a tale for winter and was interrupted when he had got no farther than the very promising beginning "There was a man dwelt by a churchyard." The literary ghost story is hardly older than Poe, but has built up a very respectable volume of literature. And at some time in the nineteenth century, which was so mechanically minded that the mere existence of the supernatural seemed ridiculous, writers began exploring the humorous possibilities of magic in the modern world; for instance, F. Anstey's *The Brass Bottle,* about a London architect who releases a jinni from a bottle and cannot control the jinni's gratitude, which takes the form of caravans of treasure in a London street and Oriental dancing girls in a London flat. These two strains combine in what is now called fantasy fiction. This has its own devotees, but they are less numerous and less vocal than those of science fiction. During the early 1940's there was a magazine, still lamented by some of us, called at first *Unknown* and later *Unknown Worlds,* which was devoted to fantasy fiction, but it failed for lack of support. Another, which began life as *The Magazine of Fantasy*, has fared better, but decided at the second issue to change its name to the cumbersome *Magazine of Fantasy and Science Fiction*, and its editor tells me it sells better when the cover features a space-ship or something suggestive of science. Even *Unknown* frequently pretended to call in science; there was a fine story about a race who were, essentially, descendants of the were-wolves of legend, but the author felt it necessary to throw in a little scientific hocus-pocus about genes and mutations. Conversely, readers will apparently accept anything if they are told it is scientific. There is a whole group of highly and deservedly popular stories taking place about A.D. 2000, less than fifty years from now, in which robots have reached such a high point of development that you can buy a mechanical duplicate of yourself and send it to take your place at a dull party, and no one will know the difference. Personally, I find it easier to believe that some day I may find a bottle with a jinni in it, but robots are scientific.

The reason both for this insistence on science, and for the sudden explosion of interest in science fiction, is pretty certainly the atom bomb. For one thing, science fiction gained credit because it had predicted the bomb. Indeed, an atomic bomb, and a world ruined by uncontrolled chain reaction, were used at least as early as the novel *Last and First Men* by the Englishman Olaf Stapledon, which appeared in 1930. More dramatically, during World War II a story appeared in *Astounding Science Fiction* which caused an investigation by the FBI, who were unable to believe that there had not been a leak in the laboratory where the bomb was being secretly manufactured, a fact which turned the spotlight on science fiction as a prophet. More important than prophecy, however, was the bomb itself. Besides the appalling physical force displayed, there was the immense political consequence, the ending of the greatest war in history—and the background question of what the other political consequences would be. There was the possibility of mutations among the children born to people who had been affected by the radiations. A favorite theme of science fiction had long been the coming of Superman, the idea that Homo sapiens is destined to be superseded by some sort of Homo superior; here was a way in which Homo superior might be born. The bomb *was* the jinni in the bottle; he was out of the bottle now, and the man in the street did not know whether to regard him as a savior or a destroyer. The possibilities of science seemed limitless; he turned to science fiction to find out what they were.

This brief account of the development of science fiction indicates why it is so difficult to define it, and for our purposes probably serves better than a formal definition. My own best effort at such a definition would be "Science fiction is fiction based upon some imagined development of science, or upon the extrapolation of a tendency in society," which sounds sufficiently broad and even vague; but even that does not cover everything with which we shall have to deal, for we shall be discussing whatever is marketed and consumed as science fiction, in print, movies, television, radio, and comic strips, whether a purist would classify it as science fiction or not.

CHAPTER TWO

Space Operas, Mad Scientists, and Bug-Eyed Monsters

As I TRIED TO INDICATE in the last chapter, the phrase "science fiction" covers at least as wide a range as that of "detective story," which notoriously ranges from stories of almost pure adventure, sometimes with overtones of brutality or farce, through chess problems and novels of suspense, to the theology of Chesterton or the sometimes excessive human interest of Dorothy Sayers. Science fiction includes all these types and more besides; but those who have not read much are apt to think of it as dealing only with pure and rather lowbrow adventure; especially since this is, almost of necessity, the stock in trade of those forms of which one becomes aware without ever reading science fiction at all—the movies, television, and comic strips. I said earlier that most people outside the circle of fans, if asked to define science fiction, would probably say, "Oh, science fiction is *The War of the Worlds* and *Twenty Thousand Leagues under the Sea,* and things of that sort." I think this is true of the older generation; but on reflection, I am afraid that a good many of the younger ones would say, "Science fiction is *Bride of Frankenstein* and *Gruesome Science Comics,* and things of that sort." Certainly, I could not blame anyone who did. Such things are less common now, but I can remember when almost the only current science fiction to be had was in pulp-paper magazines whose covers usually featured a tentacular creature menacing a girl wearing a space-helmet and not much of anything else. It is not surprising if these, and the more lurid movie posters, bulk large in the popular conception.

The magazine covers, actually, were little more than a come-on, analogous to the movies' rechristening Barrie's *The Admirable Crichton* with the name *Male and Female.* It was an attempt, if not admirable at least understandable, to lure the customers inside by any means; who knows, some of them might like what they found even if it wasn't what they expected. The actual amount of sadism in science fiction is negligible, far less than in detective stories, and there is in general no more sex than in Westerns; Westerns usually provide some sort of love interest to enhance the happy ending, whereas science fiction can get along quite happily without any love interest, and, indeed, without the happy ending; it is quite likely to end with the ultimate victory of the Martians or the robots (although this occurs chiefly in the more intellectual types which we shall consider later). As for the tentacular creatures, it is true that in the early days science fiction made some use of fearsome

creatures from other planets that merely gave fresh shape
to the old thrill of danger; but nowadays in the letters to the
editor of a science-fiction magazine you will often come
across the word BEM. This is a recognized abbreviation for
Bug-Eyed Monster, and its popularity is a fair indication
that the Bug-Eyed Monsters themselves have been pretty
completely outgrown—in books and magazines, at least.
(We shall meet them again in a moment.)

The movie posters, on the other hand, were usually a
fairly faithful representation of the high spots—or the low
spots—of the film. Until recently, science-fiction movies
have been almost entirely confined to variants of the Frankenstein theme, the vivification of dead tissue and manipula-
tion of living. There were occasional stories of the transfer
(most often only threatened) of a brain from one body to
another—usually, in these, the heroine was strapped to an
operating table amid a great parade of flashing electric arcs
and retorts bubbling like juke boxes, and the hero burst in
and rescued her at the last moment—but the most popular
was the actual subject of *Frankenstein,* the creation and ani-
mation of a humanoid monster. This sort of thing has a
respectable ancestry, going back through Wells' *Island of
Doctor Moreau,* where the doctor, for reasons that were
never apparent, attempted by a series of vivisections to turn
animals into humans, to Mary Shelley's *Frankenstein* itself,
which has given a word (generally misused) to the language. In the movies, of course, there was none of the

symbolic nemesis of creating a monster which one cannot control, still less of the sympathy which, almost against her will, Mary Shelley evoked for the monster she had set out to make purely horrible. The purpose of the movie monsters is simply to make our flesh creep; their emphasis is on the morbid. Undoubtedly, in the Brides and Sons of monsters and mummies that appear so conspicuously in the titles, there is more than a suggestion of necrophilia. But the early crop of movie monsters had no more real pretension to belong to science fiction than did the werewolves, cat-people, mummies, and zombies with whom, in the innumerable sequels and series, they sometimes got together. I am reminded of a young lady of my acquaintance who in her schooldays was a fan for horror literature generally; she once drew a map of an imaginary island whose various regions were labeled Vampires, Werewolves, Ghouls, etc. One valley was labeled Mad Scientists. That is the place for them.

Since the boom in science fiction, the movies have begun making pictures which are emotionally infinitely above the sexual horror of the monster cycles; intellectually, however, they have not often been able to get above the level of pure excitement—if it excites you. For example, one of the earliest and best-received, and indeed one of the best, was called *The Thing*. I arrived at *The Thing* a few minutes after it had begun, and almost at once was struck by the similarity in its general situation to that in a story by John

Campbell called "Who Goes There?" since in each there is a scientific expedition in the Antarctic, besieged by some sort of extraterrestrial monster. There, it seemed to me, the resemblance ended. When the screen credits came on I was quite astonished to find that the movie was based on the story. The creature in "Who Goes There?" was (quite credibly, on the pseudoscientific or parascientific terms of science fiction) a being which could on contact destroy the life, and assume not only the form but the intellectual characteristics, of any living thing. If it attacked a sled dog, it would appear to be that dog; if it attacked one of the scientists, it would appear to be the scientist, would possess his knowledge and his mannerisms; and it was divisible, like an amoeba, so that it could take over all the dogs and men—if not checked, all the life on earth. One or more of the men may have been already attacked and controlled by the creature; how are they to tell which of them is still human? That is a situation which provides not only suspense, but a real intellectual problem. In the movie version, perhaps inevitably, the creature has become a menace which, though more powerful and less vulnerable, was not, as a menace, essentially different from a gorilla. It was described as a sentient vegetable; this, if it had any reason beyond a desire for novelty, seems to have been intended to explain the creature's complete lack of the milk of human kindness, but then it was possessed of a universal ferocity which is equally foreign to the general nature of vegetables; it was highly carnivorous; it was not to be injured by most weapons, but was found to be capable of being electrocuted; in appearance, in the brief glimpse which was fortunately all that the audience was allowed, it was distinctly anthropoid. In short, a Bug-Eyed Monster.

And the science fiction in movies, to date, has dealt almost entirely with Bug-Eyed Monsters. There have been one or two exceptions, notably a fine semi-documentary of the first flight to the moon, which with great ingenuity produced effects like that of weightlessness during the travel in the space-ship, outside the earth's gravitational sphere. But that sort of appeal can be made only once or twice, and in general the movies have only looked for new types of menace, things from outer space and creatures from black lagoons, prehistoric survivals or mutated ants, occasionally varied by global disasters. These give opportunity for trick photography, and the crudest sort of excitement, and are all right if you like that sort of thing.

The other predominantly nonverbal types of science fiction, television and comic strips, are even worse. These are frankly aimed at the juvenile audience—at the sort of small fry who a year ago were wearing cowboy suits, and whom the manufacturers are now trying to get into space outfits. Indeed, there is a story of one small boy who appeared one day in the latest juvenile-outfitters' version of a space-suit, but who had not mastered the idiom that goes with it, and who greeted a friend with, "Put 'er thar, you ornery old horned toad!" "Huh!" said the friend, "blast off, spaceboy! Your jets are choked. You're talking Western, not space!" "I," said the first small boy, with considerable presence of mind, "am from West *Mars*." His confusion was understandable, for almost any standard television science-fiction drama could be translated into a Western, and vice versa: substitute a space-ship for Old Paint, a ray-gun for a six-shooter, and smugglers of the rare Martian drug *whadya-callit* for cattle-rustlers, and there you are. In one case, indeed, the change-over is not quite complete; television's Captain Video has, himself, a sort of time-traveling television set, known as a "remote carrier beam." When things are quiet in the future world that he inhabits, he is apt to tune in on the Old West, and he and his Space Rangers are treated to five or ten minutes of old Western pictures. Even the name for a certain type of stock Western, "horse opera," has been adapted, and corresponding pieces of science fiction, on the screen or off, are known as "space operas." Science fiction on TV is still at the space opera level.

Science-fiction comics need not keep us much longer. As nearly everyone must be aware by now, a distinction must be made between comic strips, which appear in newspapers, and comic books, which are sold separately on the stands. It is the comic books which are widely and rightly condemned for their sensationalism and sadism. A number of comic books have the word "Science" in the title; they are,

as might be expected, chambers of horrors worse than the
worst of the movie monster-cycles. They have of course no
real claim to be science fiction at all; their creators have
merely picked up the idea of Mad Scientists, along
with ghosts, ghouls, and gangsters, as purveyors of
cruelty.

With comic strips the case is different. The two oldest,
Buck Rogers and *Flash Gordon*, are of the space opera type,
with assorted monsters and menaces, and superweapons to
correspond. Neither of them ever made much pretense to
plausibility, to say nothing of scientific accuracy. Buck
Rogers got a knock on the head and woke up in the twenty-
fifth century; Flash Gordon, described as "a Yale athlete"
(presumably of the future, but the not very distant future,
to judge by the clothes he wore on earth, before changing to
the more flamboyant fashions of space) was engaged in a
voyage by space-ship when his ship was first attracted to the
unknown planet Mongo, and after that he managed without any difficulty to travel to various other extrasolar planets.
Buck Rogers' relations to women were of the true, austere
Western Western pattern, in which women were of little
use except as hostages, and possibly as rewards for the hero
in a future that would always be postponed until after the
end of the story; Flash Gordon had a beloved who got considerably more involved in his adventures, and also frequently encountered remarkably humanoid temptresses on
every planet that he visited; but the appeal of both these

characters has been that of the hero menaced with danger and overcoming it.

For a time these two were eclipsed by *Superman* (now, thank heaven, past its peak), a flagrant piece of wish fulfillment. *Superman* introduced itself to the world with a "scientific" explanation to the effect that ants can pull so many times their own weight, fleas can jump so many times their own height, so why not man? The answer is that he can on the planet Krypton, where Superman was born. But Superman soon leaves mere ants and bees far behind. His physical powers are literally unlimited; he can jump (one might as well say fly) as high and as far, run as far and as fast, perform as great a feat of strength, as the story calls for, and he is completely invulnerable, to mention only some of his attributes. To help the reader in the self-identification which is the only attraction of this sort of story, Superman usually appears as an exceptionally quiet, almost timid newspaperman; putting on a pair of glasses and an ordinary shirt and jacket are an impenetrable disguise for his approximately eighty-inch chest and corresponding shoulders, which must be rendered more bulky by the tights and cloak which he wears underneath, ready for the moment when he reveals himself. I say apparently because there have been times when the way he effected the change back to the street clothes which he had left, perhaps, on the other side of the world, has been even more mysterious than the lifting power of ants. But, somewhat like the dinosaurs who fell victim to their own size, or like a heavyweight champion who can find no challengers, Superman has been falling a victim to his own invulnerability, since no issue is ever in doubt for him. His creator had forgotten, but even the most juvenile public has been discovering, what was known so well to the myth-makers, that the hero must be invulnerable except in the heel, like Achilles, or the navel, like Ferrau, or the place where the leaf clung to his back, like Siegfried, or except to mistletoe, like Baldur; otherwise there is no story.

As I said at the beginning of the chapter, these predominantly nonverbal forms of science fiction probably contribute a good deal to the impression of it which is held by the public at large, which is unfortunate, for they represent it at its lowest level. It must be admitted that there is a good deal at much the same level in actual written science fiction. I have seen at least one interplanetary cowboys-and-Indians opus in a hard-cover book, whose jacket actually proclaimed it a "space opera"; the phrase is constantly used in discussions, and everybody knows the type it describes, and besides the interplanetary cowboys-and-Indians, there are interplanetary Prisoner-of-Zenda romances, novels of intrigue and cloak-and-ray-gun adventure where the countries, instead of Ruritania and Graustark, are called Mars and Venus—or, if the author is feeling really grandiloquent, Aldebaran and Altair, or Andromeda and the Magellanic Clouds.

At the best, this sort of stuff is pretty cheap candy, and, if only because adventures of this sort require enemies, it is apt to be heavily charged with xenophobia and chauvinism, sometimes a chauvinism of the human race, sometimes only a part of it. Buck Rogers' first adventure in the twenty-fifth century was to deliver the white race, whom he found groaning under the tyranny of the conquering Mongols. And everyone has heard the story of the hysteria produced by Orson Welles' broadcast of *The War of the Worlds*, when people took to the hills to escape the invading Martians, who had landed near Princeton. In too many of these stories the moral seems to be that a stranger is necessarily an enemy. And most of the space operas (which, I must emphasize, are merely the lowest level of science fiction) are not even good adventure stories, because the rules of the game are not laid down clearly. In a good adventure story the hero should overcome his difficulties by the use of strength, ingenuity, and courage, in believable and comprehensible amounts. In too many space operas, when the hero gets in a tight place he hastily invents a new death ray, or a new method of tying kinks in space, to the accompaniment of a good deal of scientific double-talk. And, just as in *Jack the Giant-Killer* each giant must have one more head than the last, to keep up some sort of suspense and climax, so after each success the hero must face bigger odds, until he is inventing his way in swathes out of entire galaxies. It does not take long to recognize that such a hero is really as monotonously indestructible as Superman. The fairy tales have always known that you could give a hero three wishes, but not an infinite number; if you did that, they must be connected with something that could be lost, like Aladdin's lamp. As soon as the reader perceives that there is no real limitation on the hero's power, he is bound to begin to lose interest.

Some of the younger readers may read space operas, as they undoubtedly read *Superman*, not so much for excitement as for wish-fulfillment, in the same way that they earlier read fairy tales, dreaming of a death ray instead of in invincible sword, a space-ship instead of seven-league boots, an ability to invent a way out of any situation instead of three wishes. There is of course one important difference in the quality of the wish fulfillment: the child dreaming of fairy gifts knows he is dreaming of impossibilities; the boy dreaming of the wonders of science believes, or half-believes, that they might actually come into existence in the future. This is something that possibly cuts both ways, but is probably more good than bad. Psychologists say that wish-fantasies are healthier if they are concerned with the possible; and though I cannot help feeling that it is better to make believe in an invincible sword than to hope for a death ray, still the boy who dreams of space-ships may grow up to join a rocket society and do something to bring his dreams nearer reality.

A more serious objection to space operas is that, being

read usually at a formative age, and being told with a considerable pretense of verisimilitude, they tend to encourage the uncritical awe of science which is one of the defects of our time. Too many of us are inclined to personify Science as a personal demigod, and to believe, according to our temperament, either in Science the Invincible Destroyer, or in Science the Invincible Savior, a force which will bring us to destruction or salvation without our having anything to say or do in the matter. This attitude is an unhealthy one, and it is probably encouraged by space operas.

But it is easy to exaggerate this sort of influence from books. I have always been doubtful of La Rochefoucauld's dictum that few people would fall in love if they had not read about it; and I am still more doubtful of the proposition, which one sometimes sees implied, that if they did not read about it, young people would not think of killing their enemies. In any case, the effects of which I have been speaking are confined to the lowest type of science fiction, which I have called "space operas." The higher types of science fiction, which we shall consider in the later chapters, are in general remarkably free from chauvinism, and give a clear idea of what is scientifically possible and what is not. And the consumption of space operas, for anyone of any intelligence, is self-limiting; nobody can go on eating cheap candy forever. The difference between science fiction and other comparable types of light reading, such as Westerns and detective stories, is that when the interest in the latter is outgrown they have nothing beyond to offer; whereas the reader of space operas, who usually begins young, is likely as he matures to be led on to the more intelligent types of science fiction, which deal with scientific problems and with social and even philosophical ideas.

CHAPTER THREE

Scientific Science Fiction

To CLEAR THE GROUND, it has been necessary to spend some time on various aspects of science fiction which are considerably more fictitious than scientific, and the reader may be wondering whether there is any science in the mixture at all, even in the proportion of one rabbit to one horse, as in the legendary stew. In most science fiction there is plenty, and has been from the beginning. The earliest science to be used as a base was naturally physics, the source of most of what we think of as scientific inventions, whether of today or tomorrow—and perhaps one should add astronomy, which gives us the mostly discouraging facts about actual conditions on the other planets of our system. Many of the fans of the early magazines were high-school kids with their physics both up-to-date and fresh in their heads, who were quick to write in and point out errors; and many of the early writers were physicists and engineers, delighted to have found a profitable sideline. The physicists even developed one type of story which stands at the opposite pole from the space opera, and like it has received a nickname of its own, the gadget story. The gadget story resembles a certain type of detective story, in being based on a single curious fact. You get hold of an odd fact, for instance that glass in water is practically invisible, and write a murder story about a glass knife hidden in a vase of water. Similarly, you get hold of an odd fact in physics, perhaps something about the peculiar properties of a certain substance, and invent a story in which there is a problem that can be solved only by a substance with those properties. These stories usually ended with somebody saying, "As long ago as the twentieth century it was known that—" and coming up with the odd fact. Pure gadget stories are growing infrequent, but the device is still used as part of longer stories. In Fletcher Pratt's recent collaboration *The Petrified Planet* one of his characters brings up a substance, teflon, which, if it could only be made transparent instead of translucent, could be made into space-suits to be used on a planet whose atmosphere is mostly fluorine gas; the character describes its properties, and adds, "Known since the twentieth century."

It is typical of present-day writing that the problem of the space-suits is incidental to the main story. In a true gadget story, where the whole solution and climax is presented by the fact "known since the middle of the twentieth century," I always mutter, "Not to me it wasn't." I may as well confess, what will be obvious to anyone who reads this, that I am a living proof that it is not necessary to have either a scientific education or a scientific cast of mind to enjoy science fiction, and I avoid gadget stories as I avoid space operas.

But even a quite unscientific person like myself can enjoy the chess-problem type of story, that in which the author provides himself with a rigid hypothesis and develops its consequences with all the scientific accuracy at his command. This type, since it seems to be almost unknown outside the circle of fans, though very common and popular in that circle, is worth describing and analyzing at some length. A recent and admirable example is Hal Clements' novel *Mission of Gravity*. The scene of this is a remote planet named Mesklin. The point of view is alternately that of an Earthman in a space-ship, and that of one of the intelligent natives of the planet—who turn out to be caterpillar-like creatures a foot or so in length. As the story progresses, we learn that the climate is intensely cold; that the gravity near the equator, where the story opens, is about three times that of Earth, but increases fantastically as we approach the poles; that the planet's rate of spin upon its axis—that is, its day—is about eighteen minutes; that it is largely covered with seas of frozen methane. None of these facts is stated directly; we are allowed to pick them up for ourselves. For instance, there is a reference to storms that last hundreds of days, which should give food for thought; later the Mesklinite character reflects that the Earthman in the ship lives an oddly periodic existence, since he appears to sleep regularly for periods of twenty-five to thirty days. That is the way in which we learn about the planet's rate of spin, and the other conditions are indicated in similar fashion. The story is concerned with a Mesklinite expedition to the polar region. The Earth ship has landed near Mesklin's equator, and has sent out an exploratory rocket under remote control to collect scientific data; the rocket, through some flaw in its mechanism, is grounded near one of the poles of Mesklin, where the gravity is so great that no earthly creature could survive; the captain of the space-ship induces a Mesklinite sea captain to undertake the voyage to the pole and recover the data in the rocket. (The Mesklinite civilization is hardly above the level of barbarism, but the Earthman supplies a radio for communication.) There are minor adventures, encounters with hostile tribes and so on; there are also hints that the Mesklinite has designs of his own, but these actually amount to very little. The interest of the book, and it is great, is the way in which the author has worked out the consequences of his planet's conditions, their effect on weather, on architecture, on the arts of war and of navigation, on the minds of its inhabitants.

Anyone can perceive, and take pleasure in perceiving, that these are ingenious and logical; an unscientific person like myself would not realize how interdependent they are if he had not read an article by the author in *Astounding Science-Fiction* magazine, describing how he worked out his conditions. It began, Hal Clements says, when he read some work on the orbit of the binary star 61 Cygni, which

proved that one of the two suns which make up this star
was accompanied by a planet. By doing a good deal of
astronomical reasoning of his own on the data given, Mr.
Clements decided that the gravity of this planet must be
about three hundred times that of Earth. Now, what sort
of story, introducing an Earthman character, could be laid
on this planet? None at all, so far; any earthly animal would
be crushed by its own weight. But gravity is counteracted
by centrifugal force; by giving his planet a fast enough spin
he could bring down the gravity on the equator to a mere
three times that of Earth—hence the short day on Mesklin.
This spin would cause Mesklin to be extremely flattened,
more like two saucers set rim to rim than a sphere, and, he
says, "to be perfectly frank, I don't know the exact value
of the polar gravity; the planet is so oblate that the usual
rule for spheres, to the effect that one may consider all the
mass concentrated in the center for purposes of computing
gravity, would not even be a good approximation if this
world were of uniform density. Having it so greatly concentrated helps a good deal, and I don't think the rough figure
of a little under seven hundred Earth gravities that I used
in the story is too far out." There is also the matter of its
temperature. Given the radius of its orbit and the size of its
sun—this is still the dark star in 61 Cygni—he calculates
that the temperature will be around minus one hundred
and seventy degrees centigrade most of the time. "Presumably any life-form at all analogous to our own will have to

consist largely of some substance which will remain liquid in its home planet's temperature range," and after considering other desiderata he decided that it had better be methane. And so on, working out in detail the consequences of the conditions he has laid down. In all this, it will be observed, the presence of the space-ship from Earth is the only condition contrary to present fact, or rather to present possibility; all the other hypotheses are either probable or at least possible. The dark star in 61 Cygni may *be* Mesklin!

Of this, Clements modestly says, "The fun lies in treating the whole thing as a game. I've been playing the game since I was a child, so the rules must be quite simple. They are: for the reader of a science fiction story, they consist in finding as many as possible of the author's statements or implications which conflict with the facts as science currently understands them." To which one reviewer (Clyde Beck, in *Science Fiction Advertiser*) appends, "There must be some fans left still who remember the time when a writer was expected to play according to these rules, and the letter columns of the magazines constituted a sort of running score sheet. A great game—you couldn't lose really. If the author made no errors you could detect, it was a swell story. If he did, sure the story was lousy, but you had the fun of catching him out." This game, obviously, is one for specialists; but even as a spectator sport it has its pleasures.

Of course, not all science fiction consists of chess problems, nor is it based only on physics and astronomy. It might almost be said that every science, exact or inexact, has been made the basis for a piece of science fiction—or if not, it soon will be, at the rate we are going. In fact, *Mission of Gravity* contains biology as well as physics; the caterpillar-like Mesklinites are built low to the ground to enable them to withstand Mesklin's gravity and are equipped with many pairs of pincers to enable them to hold on in Mesklin's gales, which in itself shows that science fiction's biology is not entirely a matter of bug-eyed monsters and unexplained mutants.

If biology has not produced as much serious science fiction as physics, it has not done badly. That chicken's heart which Dr. Alexis Carrel kept alive for so long in a saline solution has produced a number of good, well-documented menaces; and on a more philosophical level there are, for example, the works of Olaf Stapledon, one of the earliest and most imaginative writers of the era of modern science fiction. His best works, almost without plot in the conventional sense, are vast pageants of the ways in which life expresses itself, extended over galaxies and millennia. His *Last and First Men* shows humanity, Earth's Homo sapiens, again and again thrown back by some cataclysm almost to the level of the ape, and always rising, to control evolution, to create artificial brains mightier than any skull could hold, to develop telepathy through time as well as through space, and to face final extinction from an explosion of the sun. His *Star Maker* shows life spread through innumerable planetary systems, in races starfish-shaped and monopod, arachnoid and ichthyoid, its physical form always determined by its evolution and the conditions of its planet, its spirit always recognizably the same; and always developing the telepathy or symbiosis which for him is evidently the symbol of something greater even than the brotherhood of Man: the brotherhood of Life. Some of these themes—controlled evolution, artificial brains, and interplanetary symbiosis—have often appeared since (and possibly earlier; as I said before, science-fiction writers tend to build upon one another's ideas, and it is often impossible to tell where an idea originated; also, of course, the same science is likely to suggest the same ideas independently).

Pure mathematics may seem unpromising, but it has given us not only the paradoxes of time-travel, but the equations of Einstein, which science fiction has sometimes used instead of agreeing to ignore. There have been stories, for instance, of space-ships headed for stars which, even at the speed of light, are centuries away, so that entire generations are born and die who know no other world than the ship. There has been a story which applies Einstein's statement that time proceeds at a different rate in a body traveling at the speed of light; the crew in that story must resign themselves to the fate of the inhabitants of the villages of Germelshausen or Brigadoon, coming back to Earth at the end of each voyage to find their friends all dead and a new generation grown up. There are stories based on the Möbius strip, one of the elementary figures in the science of topology. You no doubt know that this is produced by taking a strip of paper, giving one end a half-turn, and pasting the ends together, so that a pencil mark drawn down the middle will come back to itself after traversing all of both sides—or of the only side there is. That has given rise to at least two stories that I can recall offhand, one about a contract to paint the inside of a strip which involved painting the whole, the other a story in which the hero was trapped in a tunnel patterned after the Möbius strip—an excellent example, by the way, of the difference between the gadget story and the adventure story.

It is not only the exact sciences that serve as a basis for science fiction. I shall speak in the next chapter of the sort of science fiction that is based on changes in society; but here it may be said that science fiction may be based not merely on the less exact sciences, such as psychology and sociology, but on what may be called para-science, and even on systems of thought which can hardly be called science at all. Recently there has been a large and growing group of stories based on such matters as have formed the experiments of Dr. Rhine at Duke University—extra-sensory perception, first of all, or the ability, as it is illustrated in Dr. Rhine's experiments, to call the denominations of unseen cards with a significantly greater degree of correctness than the laws of probability would allow. This is abbreviated to ESP, and the word "esper," meaning a person who possesses this capacity, may well follow "space-warp" into the dictionaries. There is also precognition—some people, or so it appears, can predict what the order of Dr. Rhine's cards will be after they have been shuffled; and, somewhat as with time-travel, but more urgently because more probably a scientific fact, this leads to a philosophic question: Is the future fixed? Some investigators incline to think precognition is consistent with the existence of a number of possible futures with varying degrees of probability; and there has been a story about this too. Then there is telekinesis, or the ability to move objects at a distance—in Dr. Rhine's experiments, and in the pragmatic belief of generations of gamblers, the ability to control the fall of dice by wishing. And of course there is telepathy itself, which was for many people a subject of belief long before it became an object of scientific study. There is also teleportation, or the ability to transport oneself through space—whose existence, under the name "levitation," was, like telepathy, testified to by numerous respectable witnesses on various occasions in the past. All these paranormal phenomena, grouped together under the name "the psi factor," are as familiar to the readers of science fiction as space-ships and space-warps.

Toynbee's theory of history has been the basis of science fiction. So has Korzybski's work on non-Aristotelian semantics. I do not believe I have ever seen a story based wholly on Douglas' economic theory of social credit, but I recall a story where the society in which it occurred was organized on that basis, which came out clearly enough for anyone to pick up the principles. Any set of ideas that is in the air, if only in the rarefied air of a few specialists, is likely to become the basis of science fiction—and, in the more intelligent magazines, of lively discussion in the correspondence columns.

At the end of the last chapter I said that when people grow tired of cheap candy, science fiction has more nourishing food. I am always suspicious of people who maintain that stamp-collecting is fine because it teaches geography, and I should be the last to claim that I read science fiction in order to learn science; yet I have learned a certain amount, quite painlessly. I remember that as a schoolboy, when I first encountered discussions of life on other planets, I used to object, "But it might be so different that it's no use even to speculate about it! Life on another planet might have no more in common with ours than the quality of pinkness has with a triangle," and a year's course in college physics and another in college biology (I will not say what college, to spare the embarrassment of the teachers who gave me passing grades) left me still of that opinion. Simply by reading science fiction, I have learned something about the conditions under which life is possible in this universe, even to carbon and silicon as the two systems of molecules which could form a vehicle for it, and I have some idea of the possible range of variations that living matter could show. I have strayed over into the correspondence columns and even into the factual articles which appear in the better magazines, and am almost ready to acknowledge that science itself, provided you do not have to cut up frogs or perform experiments in physics (which always seemed to involve simple arithmetic, something I cannot do) can be interesting. My type of mind has led me to get my Toynbee and Korzybski direct, but I imagine that there are many readers who are getting something of them, as I am of elementary science, through the medium of science fiction.

They are also becoming acquainted with the scientific attitude—the willingness to entertain any hypothesis, combined with an insistence on logic and on verification if possible. The attitude of these really scientific stories is almost entirely healthy. Most scientists are naturally liberal in such matters as racial equality; it is pleasant, and in a way amusing, to see how often they put in a plug for it, by making a President of the United States or some other important figure a Negro. This is the sort of thing which I often feel does more good than argument.

Gutta cavat lapidem, non vi, sed saepe cadendo—
"Water hollows the stone by striking not hard, but
   often."

A boy who has absorbed a prejudice against Negroes from the atmosphere of his surroundings may be led out of it by finding, in literature, another atmosphere, one where racial equality is taken for granted. And just as, in genuinely scientific science fiction, the natural equalitarianism of most scientists is an antidote to what I have called the chauvinism of Earth in the space opera, so, especially in the chess-problem stories, the insistence on the actual limitations of science is some corrective to the tendency (not confined to space opera, but a general human tendency) to regard science as either the Irresistible Destroyer or the Omnipotent Savior.

There is one objection that may be made to this. In some, though by no means all, of the stories of this type there is a tendency to introduce what that dyspeptic old fascist Thomas Carlyle would call the Hero as Scientist. The scientist-character emerges as the Savior of Society—and one remembers that that title was given to Napoleon III after he had overthrown the Second Republic and made himself emperor. It is natural for anyone, especially a writer with some scientific training, to feel that he knows what ought to be done if only people would listen, and gratifying to make them do so when they are his characters. And it is natural to assume that in a crisis the peculiar gifts of a scientist would lead to his taking command so long as a unified command was necessary. Some stories, however, go farther than this, and present a scientist, or a group of scientists, keeping the people in the dark, or actually deceiving them for their own good. There is no need to enlarge on how dangerous a principle this is.

But there are not many instances of this, and science

fiction as a body of writing is not likely to be an effective instrument of propaganda, whether for rule by scientists or anything else, if only because its writers are men of such widely divergent opinions, as you will see if you will look at the correspondence columns of the more intelligent magazines. There are even writers of science fiction who deeply fear and hate science, Ray Bradbury for instance, and writers of utopias in reverse such as Aldous Huxley, though these scarcely belong among the writers of scientific fiction; we shall meet them in the next chapter. But even without this, there is room among the more scientific writers for all kinds of viewpoints. The one thing they have in common is what I have called the scientific attitude. The solider sort of science fiction is the earliest, and for many of the more academically minded of us, the only place where we encounter the scientific attitude. And it is an acquaintance worth making, since for good or ill it is one of the great formative facts of our world.

CHAPTER FOUR

Speculative Science Fiction

IN FRANK STOCKTON'S STORY "The Griffin and the Minor Canon," the griffin (who by the way is not a griffin at all, if Stockton's description is correct, but a wyvern) makes the profoundly philosophic remark, "If some things were different, other things would be otherwise." Science fiction as we have been discussing it so far has been largely an application of this principle. But there is another aspect of science fiction which reminds one rather of the even more philosophic charlady in the *Punch* drawing, who sighed, "Things might be so different if they wasn't as they is." This is the science fiction which takes as its starting-point the inexact sciences—psychology, anthropology, and sociology—and constructs whole new races and societies.

Pure utopian novels, those which draw up blueprints for an ideal society, like Bellamy's *Looking Backward* and William Morris' *News from Nowhere*, may have a scientific framework telling how the new society came about if the scene is Earth, or how the hero got there if it is the future or a distant planet, but they are hardly science fiction in the true sense. Even so, I imagine Wells would have claimed the title for his tedious utopian novels such as *Men like Gods,* in which everybody has miraculously become wise and altruistic. Wells, writing in the fools'-gold age of the twenties, was I believe the last of the utopians. Men like Bellamy and William Morris in the nineteenth century, and Wells as late as the nineteen-twenties, thought they could portray a society so obviously attractive that people would try to build one like it. Today the best hope of writers like Aldous Huxley (in *Brave New World*) and George Orwell (in *1984*) is that if they portray vividly enough the hideous future we are shaping for ourselves, we may take the warning and find a way to avoid it. We can get a grim picture of the progress of civilization in the last quarter-century by contrasting Wells' idea of a possible future, the idyllic near-anarchy of *Men like Gods,* with the iron regimentation of Orwell's *1984.* Or, to take even a pessimist of the twenties, in 1928 E. M. Forster wrote a fantasy called "The Machine Stops," a picture of a world in which everybody lives all alone in an underground cell, conversing with his fellows by visual telephone, and provided with all luxuries and necessities by one vast machine. The machine, built long ago, is beginning to break down here and there, and nobody knows what to do about it; but the point of the story is not the breakdown of the machine, a breakdown which may (or may not) be the ultimate salvation of mankind; it is the picture, symbolic rather than literal, of a world in which Man is so coddled by machinery that he has well-nigh lost the use of his muscles and his emotions. In contrast to the over-civilization which seemed probable to the imagination of 1928, consider a story like Stephen Vincent Benét's "By the Waters of Babylon," with its picture of a young savage visiting the awesome ruins of New York. That theme has been used so often that the fans have given it a regular name, Our Savage Descendants—except that in its most recent appearances it tends to be not about our descendants but ourselves, those of us who survive, some five years from now, living in caves and fighting off raiders with bows and arrows.

These utopias in reverse have their value as warnings of perfectly genuine dangers; though it is comforting to remember that they may not happen. One of the best of the stories about Our Savage Descendants was John Collier's *Full Circle*, about England after the next war. But that was written in 1933, when the next war was World War II. The ides of that particular March have come and gone. Jack London, too, wrote both a story of Our Savage Descendants, *The Scarlet Plague*, and a story of a Brave New World, *The Iron Heel*. Both were inspired by his bitter socialism (it would be communism today). *The Iron Heel* risks a remarkably short-range prophecy. It was written in 1907. The manuscript, which constitutes the bulk of it, is supposed to have been found some seven centuries hence, a device adopted to allow London to supply his supposed future audience with satiric footnotes on the barbarous twentieth century, but the action takes place between the years 1912 and 1932, during which time London imagines that the United States was converted to what we should call a fascist state. *The Scarlet Plague* is set in the ruins of what was once such a state—the old man who tells his grandsons of the debacle says, "2012—that was year Morgan the Fifth was appointed President of the United States by the Board of Magnates, sixty years ago"—and in reading it one cannot help feeling that London set up such a state not merely for the purpose of knocking it down, but for the pleasure of seeing it suffer. The old man tells his grandsons how the world was struck by a mysterious plague, which was highly infectious and invariably fatal; the sole survivors were the one in a thousand or so who were naturally immune. There is no reason why this should have caused absolute chaos; but, says London, "Down in our slums and laborghettos we had bred a race of barbarians, savages; and now, in the time of our calamity, they turned upon us like the wild beasts they were and destroyed us"—burning cities and shooting people for no reason whatever. The old man who is telling the story was of course one of those gifted with immunity, and he ultimately joined a group of survivors, in which, as London describes with obvious relish, the most beautiful woman in America, belonging on both sides to the billionaire aristocracy, had been made the slave of a brutal chauffeur, because he was the strongest man in the group. And now, two generations later, society has fallen to the bow-and-arrow level; it has occurred to nobody to look up, among other things, the receipt for gunpowder in one of the books that must be surviving somewhere— books housed in steel and stone libraries are remarkably hard to burn. Obviously, London was simply enjoying himself by imagining a day when muscle would count for more than money, and brutality more than brains.

Since London's books appeared we have seen, if not the world, at least whole communities ravaged by disasters as bad as the scarlet plague, and no such wholesale breakdowns into savagery. We have also lived through the period which London predicted would see the coming of American fascism; and whatever one may say of 1932, it was not the year that witnessed the triumph of privileged wealth. So we may hope that things may not be as bad as predicted by the science-fiction prophets of doom—and of course if they are not we have the prophets in part to thank for their warnings.

Of course, there are many science-fiction stories of strange societies that are neither utopias nor utopias in reverse, but simply attempts to represent society as it may become, or even as it might be "if things wasn't as they is." In the nineteenth century, when the age of invention had visibly begun, there were several not very successful attempts to visualize the society of the developed machine age. One of the unlikeliest authors of such a book was Anthony Trollope, whose *The Fixed Period*, published in 1882, is laid in 1980. Trollope's vision of the future was not inspired. A cricket match where the classic fifteen players a side has become sixteen and the bowling is done by machinery, and a steam tricycle which travels at the rate of twenty-five miles an hour, are among the more daring flights of fancy. There were other attempts to visualize the machine age, but they were not much more successful. It was easy enough to imagine a steam tricycle or a horseless carriage, but nobody, so far as I know, attempted to imagine the social consequences of the automobile.

Present-day authors are more given to social extrapolation, that favorite word. The title of one of Robert Heinlein's stories of the future, *If This Goes On* . . ., might serve for many others. These are stories which plot the curve that will be traced if some tendency in our present society increases to its logical extreme. Suppose there is an increase in the present tendency for the most intelligent parents to have the fewest children, until the bulk of the world's population is made up of morons. Suppose the emancipation of women continues until the nineteenth-century relation of the sexes is reversed. Suppose (this story was to me more horrifying, because more probable, than any picture of civilization lapsing into barbarism) there is a great increase in the present taste and license for sadism and decadence, at present represented by such things as horror comic books and women's wrestling matches. Those are the first instances that come to my mind, from books that I have read and remember; there are a thousand others. Other imaginary societies, usually placed on some other planet, are like non-Euclidean geometry, based on another set of postulates than ours. What would it be like in a society where sexual play was indulged in freely and openly, but the act of eating was both sacramental and shameful, something to be performed in private and never spoken of in polite society? What would it be like in a society where everyone acted on Thoreau's principle of civil disobedience? What would it be like if a life-span of thousands of years, or actual immortality, were possible for a few persons—and what would it be like if it were possible for everyone? Again, there are a thousand stories based on such speculations.

Such speculations can be real intellectual entertainment. They can, like the instance (from Olaf Stapledon) of the country where eating was tabu, be a revealing criticism of our own mores. And, as has been suggested by John Campbell of *Astounding Science Fiction* magazine, they can be useful trial sketches of possible futures. We live in a world where nothing is certain except change; but we can try to imagine the effects of certain changes and whether or not we like them, and then can even perhaps try to work for or against them.

Man and society are of course so closely connected that a change in the character of either one must lead to a corresponding change in the other. My last instances of alien societies had to do with changes primarily in Man; and science fiction has an abundance of books based on the question "What would it be like if there were people that weren't human?"—books which present characters that are almost human, but human with a difference, from the Superman at one end to the robot at the other. The industrious court genealogists of science fiction might trace the robot back to Talus, the iron man who was Sir Arthegall's attendant in *The Faerie Queene*. Without dragging him in, we may say that Frankenstein's monster was a real literary ancestor, though he was an android rather than a robot— but, as far as that goes, the original robots, in Capek's play, were androids. The distinction, it should be explained, is that a robot is an artificial man (usually—there have been robots, specialized in function, who were non-human in form) moved by machinery, thinking by machinery (in these days of computing machines and electronic brains), and like any machine deriving its power from an outside source. An android, on the other hand, is a synthetic man, made in a laboratory out of protoplasm, and, once made, capable of moving and living on independently. A robot, in the modern sense, may present the appearance of a machine, clanking and clumping, or he may verge on the android by being covered with synthetic flesh—and in that case it is always a dramatic moment when the flesh opens to reveal the wheels inside. I do not know who first made the distinction, which seems to be agreed on by writers and fans; it is of some value, but etymologically it should be reversed, since the original robots were not machines, but synthetic humans.

The word *robot* was coined by the Czech playwright Karel Capek, from a Slavic root meaning *to work*, and was used in his play R.U.R., which was an immediate success in several countries when it was first introduced in 1922, and which closed in four nights when it was revived in New York in 1942. Both the success and the failure are easy to understand. The novelty and the genuine melodrama carried it in the twenties; the absence of novelty and a certain naïveté in the solution were too much for it twenty years later. Since so much has stemmed from it, the play is worth considering in some detail. R.U.R. stands for Rossum's Universal Robots. Rossum was an inventor who had created a form of synthetic protoplasm and used it to make artificial workers, submissive and obedient, manufactured with no qualities except those desirable in a worker. But the robots are being constantly improved, made stronger and more intelligent so as to be able to do harder and more complex tasks. At the beginning of the play they have just been provided with a sense of pain, because without that they unconsciously injure themselves and so have to be scrapped sooner than would be caused by normal wear and tear. Also there is a doctor concerned in the manufacture who insists on constantly making the robots more and more nearly human, simply to gratify his own creative urge, or to play God, whichever you want to call it. The inevitable happens: the robots revolt and massacre the human race. The play's penultimate scene shows the last half-dozen left alive, barricaded in a room surrounded by robots. (I still remember from the original production the cold chill when one of the characters peers out at the besieging robots and says, "We shouldn't have made their faces all alike.") The humans have, they believe, the final trump—the formula for manufacturing more robots, when these wear their bodies out. But the formula has been destroyed, by an old servant-woman who thought the manufacture of robots was impious; and the last survivors are butchered, all except one old carpenter. In the last scene the robots are desperate; they are dying out, and they have been unable to find out how to manufacture their successors. But the youngest robot and robotess show that they have fallen in love; and the old carpenter says tenderly, "Go, Adam; go, Eve; the world is yours." "Evidently the doctor who was bent on making the robots human has given these the capacity to fall in love; it is to be inferred that he has also given them the capacity to get married. In the twenties we all seemed to be too delicate to ask anatomical questions, even of ourselves; the general impression was not of anything so crude as a glandular urge, but rather of the Life Force insistently finding a channel for itself—which is what I meant when I spoke of a certain naïveté in the solution.

R.U.R. is certainly oversimple: the doctor who insists on humanizing robots for no reason except that he has the power to do so is perilously close to that figure of burlesque, the Mad Scientist; the end is facile and perhaps sentimental; and yet *R.U.R.* is one of those works that have the power to set the imagination going along very deep channels. At the time of the first production I think it was generally taken as an allegory of the revolt of the downtrodden, faceless masses, with a dash of the omnipotent Life Force thrown in—two ideas that were very much in the air at that time; but there is much more in it than that. The framework, the constant improvement of the robots past the safety point, is much like the old folk tale of the Fisherman and the Carp, where the fisherman's wife asks the fairy carp for a neat cottage, a castle, a palace, and then to be God Almighty—and finds herself back in the old hut. It is the moral "Don't crowd your luck; Man should be content with enough." But more than this, the whole concept of the revolting robots is a very ancient one. It is not merely the story of Frankenstein; not merely the mediaeval Jewish legend of the Golem, the gigantic figure which could be animated but could not be controlled, until with the fearlessness of innocence a little girl pulled out the parchment bearing the name of God. It is the story of the Fisherman and the Jinni from the Bottle; it is the story of the Sorcerer's Apprentice, who stole an opportunity to read in his master's book and raised a host of spirits that he could not lay. That is, it is the fear that Man may somehow create a force which is too great for him to control—a fear that is very old and widespread, as these stories show, and one that we have more cause to fear than any generation has ever had before.

The robot in revolt is still used, notably by Ray Bradbury, a writer so skillful that he manages to be popular with nearly all science-fiction fans in spite of the fact that his approach is not merely unscientific but antiscientific. He plainly hates and fears the growing power of science, and has written at least one utopia in reverse. He attacks the advance of science in stories laid on a Mars whose gravity and atmosphere apparently are no different from those of Earth; or in stories which postulate that by A.D. 2000—less than fifty years hence—it will be possible for me to buy a robot-counterpart of myself which is so realistic that I can send it to a dull party in my place. In this story the robots take over not because they are stronger than their masters, but because being exactly like them can impersonate and replace them. This too is a recasting of an ancient myth, that of the Doppelgänger or fetch, the belief that it is possible to meet oneself face to face, and that to do so is an omen of death.

But not all robots have been in revolt. Isaac Asimov, in a book called *I, Robot*, assumed that the danger of a revolt had been foreseen and that all robots were made with built-in controls which completely inhibited them from any action that would harm a human being or from letting a human being come to harm through inaction. These "Laws of Robotics" have been taken over by a number of writers, who have produced the idea of robots as a race of pure altruists. There have been stories of worlds inhabited only by robots—worlds where Man has killed himself at last, or planets where only robots survived the wreck of the spaceship, worlds where Man has been forgotten, or is remembered with mistaken reverence as a vanished deity. Some of these stories are merely ingenious, and occasionally they are gruesome—for example, a story of a world of robots who killed a spacewrecked explorer by treating him, with the best intentions, as if he were made of metal. But most of these stories have depicted an innocent world of creatures like Man, but without Man's passions of sex and fury—in short, the Garden of Eden, the Earthly Paradise that Man is obliged again in every century to invent, or to remember.

A special type of robot or android is the Great Brain. So far as I know the first Great Brains originated in the work of Olaf Stapledon, who originated so much; but it is impossible to read everything in the field, and I may be mistaken. In Stapledon's *Last and First Men*, published in 1936, some of our inconceivably remote descendants construct huge brains made of protoplasm; recent developments in computing machines, or "electronic brains," have led to more recent stories of artificial brains. As is true of nearly all aspects of science fiction, some of these stories have been simply farcical, showing the artificial brains afflicted with calf love or some other form of intellectual growing pains.

More of these stories have made a sincere attempt to make the great brains vehicles of a superhuman wisdom, to show them giving answers that will stop war and violence. The trouble with these stories is that a stream cannot rise higher than its source. I can believe or pretend to disbelieve that we shall one day travel faster than light; but it is very difficult for me to believe that Man can make a machine that is wiser than he is—stronger, faster, more skillful and accurate, of course; but not wiser. And it is obviously difficult for a writer who is not himself a superman to give a convincing portrayal of superhuman wisdom.

And yet, surprisingly, science-fiction writers do remarkably well in portraying some form of Superman. To avoid confusion with the comic strip, I had better use the phrase commonly used in science fiction, *Homo superior*—the same genus, but a different species from Homo sapiens. In 1935, Stapledon's character Odd John, himself a type of Homo superior, declared, "Homo sapiens is a spider trying to crawl out of a basin. The higher he crawls, the steeper the hill. Sooner or later, down he goes. He can make civilization after civilization, but every time, long before he becomes really civilized, skid!" Since 1935, Homo sapiens has taken some spectacular skids, enough to lead many writers to share Stapledon's view that Homo sapiens is a creature that must be superseded; and the mutations which result from exposure to atomic radiation have given a means by which Homo superior might theoretically be produced. So we have had stories of mutants possessed of hereditary telepathy or other psi factors, and mutants who are simply Man's vast superior in intelligence.

There is one rather remarkable quality common to most present-day stories of Homo superior. It used to be assumed that Homo superior, with his greater intelligence, would come as a conqueror; there were even stories filled with what I have called the chauvinism of mere humanity representing Homo sapiens successfully resisting his superior rival. Later writers have recognized that Homo superior is more likely to be the potential savior of Homo sapiens from himself, and most likely of all to be Homo sapiens' actual victim. In the first place, any difference from mankind in general is likely to be at least at first a disadvantage, and all the more if it is potentially advantageous, like eyesight in Wells' "Kingdom of the Blind," since the bulk of humanity will probably declare a preventive war against it; and in the second place—if you had been born in a tribe of gorillas, how far would your human intelligence take you? Stapledon's Odd John was a freak, who managed to find a few other freaks like himself and to found an island colony, but this was wiped out by our society because the freaks were too humane to use the weapons they might have used. A. E. Van Vogt's Slan, in the story of that title, was a hereditary telepath, in a world where hereditary telepaths were persecuted. Wilmar H. Shiras' "In Hiding" shows a boy born with immensely superior intelligence; he is far luckier than most in his family and associates, but at least until he is grown he is forced to fantastic shifts to conceal his difference.

This widespread idea that we must look for the coming of Homo superior, and that when he comes he is likely to be persecuted is not the expression of any particular philosophy. The Christian apologist C. S. Lewis has written two science-fiction novels, which present Mars as a world where the Fall of Man never took place, and Venus as a world where the core of the story is the temptation of the Venusian Eve. He would undoubtedly say that the prevalence of this theme reflects the instinctive hunger of the soul for Christ—for a savior who is like man and yet unlike him, and who is persecuted for his unlikeness. Olaf Stapledon, on the other hand, can best be described as a mystical agnostic. His best books cover aeons and galaxies; their charactors are whole races, many of them fantastically alien in appearance—arachnoid, ichthyoid, sessile or symbiotic, even to a race of intelligent flames; but all of them felt by the reader to manifest what Stapledon can only describe as "Spirit." Stapledon would say—in effect he has said—that Homo sapiens must be superseded, and that we must rejoice in the coming of a race that can make a better use of its opportunities than we have been able to do. He would say that Homo sapiens is a minor failure on the part of the Life Force, but that this is unimportant in a universe filled with beings that manifest Spirit—and what Stapledon calls Spirit is a quality, independent of bodily form, which is what we feel to be the essence of Man at his best.

These stories of Homo superior belong to very different philosophies; but they have in common a very important lesson, one which is also being taught by many, though of course not all, of the stories of encounters with alien races. They teach that to be different is not necessarily to be evil. If we are being watched by beings from other worlds in flying saucers (and many intelligent people believe there is evidence for it) that lesson is of the highest importance. And if there are no flying saucers and we merely have to live with our fellow men on earth, it is still of the highest importance.

CHAPTER FIVE

Science Fiction and the Emotions

IN A SENSE, this whole book has been about science fiction and the emotions, since all writing must arouse emotion of some kind (if only the emotion of boredom), and it is impossible to discuss any sort of writing, from space opera to philosophic speculation, without giving the reader some idea of what his emotional response to it is likely to be. But it has seemed worth while to give the matter of the emotions some special consideration, since it is in this regard that science fiction is most often adversely criticized. One of the best of our living critics, Bernard De Voto, has called it "A form of literature which has almost completely succeeded in doing away with emotion." And in the introduction to his anthology *Stories of Tomorrow*, William Sloane says, "Not long ago a literary critic and teacher whose judgments I generally respect even to the point of reverence observed, 'Perhaps I ought not to put it quite this way, but I am afraid science fiction is here to stay. It's too bad that the most vigorous new development in writing should be so narrow in range and so special in interest.'"

I hope that the last chapter has shown that the range of

science fiction is not really so narrow; but I think that the second critic had in mind essentially the same thing as De Voto, a lack of emotional values. There is much to be said for this objection. With rare exceptions, modern science fiction is sadly deficient in characterization. Almost the only characters one remembers as characters are the misfit supermen, Stapledon's Odd John and the rest. This is easily understandable. In the nature of things, the emphasis of science fiction is on inventions and ideas rather than people. The characters, at one end of the scale, are often involved in gigantic catastrophes where they are acted on rather than acting, or else are the all-conquering heroes of the space operas, who are so handy at dreaming up new gimmicks in every difficulty that even for an adolescent no real self-identification is possible. And the authors might well protest that faced with the problem not only of telling a story but of giving a clear idea of some previously unheard-of state of things, they have no space for characterization. Still, Wells managed to differentiate his two First Men on the Moon; after thirty years and more I still remember Verne's Phineas Fogg and Captain Nemo, as well as his comic stage-Englishman and stage-American types; and it does seem as if science fiction ought to be able to produce a hero with as much personality as, for instance, Rex Stout's Archy Goodwin. In justice to the genre, it ought to be added that some of the younger writers are beginning to explore the human values of catastrophe or of pioneering on new planets, and this tendency will probably increase, along with the general trend away from the strictly scientific. Still, it must be confessed that in general science fiction is lacking in characterization and in the particular sort of emotion that goes with characterization.

But this is not the only emotion to which writing can appeal. There are many others—for example, humor, an aspect of science fiction which is much neglected. There has been little critical writing of any kind about science fiction, and so far as I have been able to find out none of it touches the humorous kind; most people who know science fiction only casually are surprised to learn that there is such a thing. Yet there are authors who specialize in it, and some of it is very funny indeed. (To me, at least; appreciation of humor is a highly personal matter.) After all, the broadest possible definition of humor is the perception of incongruity, and the essential material of science fiction is things that are incongruous in our present world. At least as long ago as *A Connecticut Yankee in King Arthur's Court* it was demonstrated that time-travel may have its humorous side. L. Sprague de Camp's hero who drops through a crack in time and averts the fall of the Roman Empire has some experiences which are very funny to us, if not to him, before he has solved his immediate financial problem by setting up a still and introducing the Romans to hard liquor. Or if the hero travels not in time but in space, he may encounter societies or races whose difference from ours may be used with humorous effect; for example, the same de Camp's story of a college fraternity on Earth initiating an exchange student who happened to be a small, intelligent dinosaur—small for a dinosaur, that is. Again, the Freudians tell us that at least one source of laughter is our pleasure in seeing the violation of a taboo; and science-fiction heroes frequently find themselves in societies whose taboos are not ours, where they unintentionally shock the natives or are shocked by them. In this connection it is noteworthy that in almost every utopia from St. Thomas More's down, the utopians have gone naked either habitually or on occasion. Or the incongruity may occur in our own world. The inventions of science fiction need not be all on the scale of space-ships and ray-guns. There has been a series of stories about a professor of mathematics, Profesor Cleanth Penn Ransome, who invents such things as a non-Euclidean basketball that changes its size and shape during a game; and another series about a man who constructs cockeyed inventions when he is drunk, and finds when he sobers up that he can no longer remember what they are meant for or how to control them.

Even when it is not deliberately humorous, science fiction above the space-opera level often has humorous touches. There is likely to be something funny about a man placed in a position where he cannot believe the evidence of his own senses, which is something that often happens in science fiction. Parenthetically, the people it happens to are often the high brass; science fiction has so many instances of top-ranking generals and military administrators who attempt to deny the existence of the hitherto unknown, or to do away with it by issuing orders, as to warrant the suspicion that most of the people qualified to write science fiction were not the type to find a home in the army. Francis, the talking mule of the movies, is an instance of this, though his stories make no claim to science.

A man who cannot believe his senses is funny only if his self-confidence is so overweening that, so to speak, he does not really believe in his own disbelief. If he could, he would be terrified, and an object for pity rather than laughter. It is always true that an incongruity may be so big as to be no laughing matter, and then it provokes pity or terror. "Man is the only animal that laughs and weeps," says Hazlitt, "since he alone perceives the difference between things as they are and as they ought to be." Science fiction appeals to both these ancient emotions—with, naturally, varying degrees of success. It must not be forgotten that the term science fiction covers almost as wide a range of merit as fiction itself. One obvious, perhaps too obvious, source of terror has been the atomic and hydrogen bombs and their consequences. For a while after Hiroshima there was such a flood of stories of the next war and its aftermath that readers finally became numbed to them and editors were obliged to call a halt. These stories were written, no doubt, partly out of real missionary zeal to warn us of what we may be preparing for ourselves. They piled on the horrors—cannibalism and pestilence, nations living underground, their whole failing organization geared to a war whose causes had been forgotten—to such an extent that the mind refused to accept them; they became as incredible as the galactic-scale slaughters of the later space operas. Or the devastation after the war became the scene of adventures that had, inescapably, a certain Robinson-Crusoe appeal. But occasionally a writer would concentrate upon a single family, and actually make an atomic war seem real. I am particularly thinking, as everyone who has read it will be thinking, of Judith Merril's story "That Only a Mother," which, in a quiet, peaceful setting, administers the full shock not of war but of the effects of radiation. That, and a few stories like it, recall the full meaning of the phrase "pity and terror," and we can only hope that they will purge our souls.

Stories of the atomic war are all too possible; there are, of course, many scientific terror-tales which are impossible so far as we know at present. It has been suggested that science fiction is the modern form of the "ghost story"—a phrase, which, as I have remarked, is as vague as "science fiction." It has been suggested, that is, that science fiction is replacing the supernatural tale of terror, and that it is more effective because, it is said, we no longer believe in ghosts, but we can believe in invaders from Mars or flying saucers from Alpha Centauri. With this I do not agree. I preserve an open mind about flying saucers—for that matter, I preserve an open mind about ghosts—but it would take something at least as effective as Orson Welles' Martian broadcast to make me frightened of a flying saucer. I should be frightened of an invasion from outer space only if I believed it was actually happening or about to happen; whereas the idea of the unquiet or vengeful dead is so deeply ancestral that it is not necessary to believe in ghosts to be frightened by them, or at least moved by them—witness the success of Irwin Shaw's antiwar play *Bury the Dead*, about the dead of World War I who refused to lie down.

In denying that science fiction is replacing the ghost story, I do not mean that it cannot touch the nerve of terror in its own way. Since I have been collecting ghost stories, as well as science fiction, since my schooldays, I can boast that my blood has a high, acquired tolerance for alcohol and curdling, it has taken so much of both; and I am happy to say that at its best, science fiction can still produce in me the authentic chill. There is Ray Bradbury's story *The Crowd*—admittedly close to fantasy, but with nothing in it contrary to possible fact—the story of a man who begins to suspect, and then to be sure, that the people who appear at an accident, pressing round the victim, staring and shutting off the air, are always the same people; he becomes sure of it only in the moment when his own accident becomes fatal. It is the theme of the ghoul. Or there is Jack Williamson's

With Folded Hands, a story of how mankind is taken over by a race of super-robots whose only wish is to serve and protect Man and keep him from harm. They do all the work; they gratify every wish; they will build you palaces if you like; they will never allow you to be harmed, or even to run the risk of being harmed. The only catch is that anything you really want to do, ski or take a second highball or strain your eyes by reading all evening, turns out to be something that might harm you, and the robots cannot allow you to do it. That is as fantastically gruesome as the end of Evelyn Waugh's *A Handful of Dust,* and is the sharper because, as sometimes happens in sleep, it is the dream that turns into the nightmare. Parenthetically, I cannot help thinking it a great pity that the author wrote a sequel explaining that it was all for the best and was the only way to keep us from blowing ourselves up with atomics. Better the race should die like men, even if madmen, then live like pussycats.

I myself find science fiction most disturbing, not when it prophesies, but when it holds the mirror, and the magnifying glass, up to the darker places of man's mind. I do not seriously believe that there will be either an invasion from outer space nor a revolt of the robots in my time. (As to the robots, we have to develop them first, and I have a good chance of living out my life in the happy *ancien régime* before they revolt.) It is not even certain that there will be an atomic war. But it is certain that I shall die. When that times comes, I hope I shall be able to forget a story based on the assumption of a perfect telepathic relationship, where one of the partners died, and the survivor discovered to his horror that the psyche remains conscious so long as any of the brain is left, but the decay of the brain causes madness and torture. Or, if you wish to be a rigorous logician, you may say that it is not certain that I shall die since science may come up with the elixir of life; but it is certain either that I shall die or that I shall live forever. How do I feel about that? Immortality, life everlasting, has always been the greatest of man's wishes and of religion's promises; but when we seriously think of living for eternity, or even for millions of years, the prospect becomes tolerable only if we assume that not only our surroundings but ourselves will be so altered as to be almost unrecognizable. Nearly every scientific utopia faces the problem of euthanasia; some stories have been written directly facing this problem: suppose that the life-span can be prolonged indefinitely, giving, practically, immortality, what happens then?

Finally, science fiction on its higher levels has the appeal of myth. This is something that has often been said, but I think it is true at a deeper level than is always realized by those who say it. It is sometimes said as if all it meant was that science fiction offers the old wish-fulfillment fantasies of the fairy tales in allotropic forms, offering the space-ship instead of the flying carpet or seven-league boots, the ray-gun instead of the invincible sword, and so on. I have spoken of this in an earlier chapter. It is true as far as it goes, but it does not go far enough to explain why anybody over the nursery level should read science fiction in preference to the Oz books. The same comparison has been made on a somewhat deeper level by Gilbert Highet in his radio program *Persons, Places, and Books,* a series of literary causeries. In one of these Professor Highet decried science fiction as a waste of time, wondering what anyone could see in it. It is plain that at this time, in the most literal sense of the words, he did not know what he was talking about; he had read little or no good modern science fiction. After his first program he was flooded with letters and books urging him to read this and that; and to his great credit he devoted a second program to the subject, acknowledging that it had more values than he had realized, and calling it the modern myth, but at the level, if I understand him, not of the fairy tale but of the hero tale. The journey to the stars, the man alone against the inhuman enemies, is our version of man against the superhuman powers. This is claiming for science fiction too much and too little. Too much, because though someday we may have a story of galactic warfare equal to the *Iliad,* or a journey to the stars which equals the *Odyssey* or even the story of Jason, we have not had them yet, and I should not like to stand on one leg until we have. It claims too little because science fiction does more than embody our aspirations; it restates certain truths which can only be stated in terms of what is here and now impossible. In the past chapters I have pointed out how science fiction has restated the themes of numerous myths and fairy tales, notably the dangers of having your wishes granted.

Some thirty years ago there appeared a series of books called the *Today and Tomorrow* series, each attempting to forecast the future along a particular line, and usually with a classical allusion in the title—*Narcissus, or the Future of Dress; Lars Porsena, or the Future of Swearing and Improper Language*. The series was inaugurated by two books, one by J. B. S. Haldane, entitled *Daedalus, or Science and the Future*; the other by Bertrand Russell, entitled *Icarus, or Science and the Future*. That illustrates what I have called the mythical aspect of science fiction. Science, not in imagination but in fact, has given us wings. Science fiction promises us that they may lead us out of the labyrinth, or may plunge us into the Icarian Sea.

CHAPTER SIX

The Future of Science Fiction

IN FICTION AT LEAST, John Collier predicted that England would lapse into savagery after World War II; Jack London, that America would be a fascist dictatorship by 1932; G. K. Chesterton (in *The Napoleon of Notting Hill*), that England in the 1990's would show no important technological advances over England in the 1890's. With the examples of three such better men before him, a forecaster may well hesitate to approach the crystal ball; but it may be possible to extrapolate tendencies and reason from analogies.

To begin with, is the present vogue a boom that will be followed by a bust? I do not think so. Its nearest analogues as popular entertainment are the detective story and the Western, the detective story especially; the rise of science fiction after World War II is like that of the detective story after World War I. Both these other forms have had a longer life than science fiction, and they are showing no signs of senile decay. Critics have often declared that the detective story has at last exhausted all the permutations and combinations of its material and is about to wither away, but it continues to flourish; and the possible range of science fiction is enormously greater than that of the detective story. A few years ago I should have said that the detective story at its worst could not be so bad as science fiction at its worst, since the detective story had to be based on some sort of reasoning process; but that was before the rise of the school of blondes and brutality. Now I should say that the detective story at its worst pulmbs depths of sadistic stupidity which are well below the worst of science fiction, while science fiction at its best has potentialities far higher than the detective story at its best. Of course, since the detective story is so much older, it has realized far more of its potentialities; admittedly, there are in existence more first-rate detective stories than first-rate science-fiction stories. But the potentialities of science fiction are higher, for several reasons. The detective story must keep us outside the mind of the criminal, and if we are not to suspect the one character whose mind is a closed book, it must keep us out of the minds of most of the other characters as well, whereas science fiction is free to present its characters in full. The detective story is almost obliged to treat violent death as mere ground for ratiocination or excitement, sometimes farcical excitement, whereas science fiction can and often does present tragic events tragically. And, as I have tried to show in the last chapter, the range of themes open to science fiction is enormously greater than that of the detective story. Since 1914 the detective story has branched out into many types, sometimes, as in the last of Dorothy Sayers' mysteries, almost forsaking the detective form for that of the pure novel. Some publishers have a set of symbols with which they mark the jackets of their books, to indicate whether a book is a chess-problem, a cops-and-robbers chase, a homicide with humor, or whatever. The same thing is already happening in science fiction. In the magazine field, the principal magazines have each their special type of contents, and their particular audience. The science fiction of the future will probably be still more sharply differentiated into adventure, farce, gadgetry, speculation, and fantasy. Of these, it may be hoped (though not too hopefully) that the general level of the adventure and farcical stories will improve, as the general public becomes aware of the concepts of science fiction. At present, the mass media—movies, television, radio, and comic strips—are obliged to stick close to the common denominator of pure adventure, because that is all the public knows about. If from these forms of entertainment the public learns to understand and accept some of the conceptions of science fiction, there may be a rise in the quality of the writing comparable to the rise in quality of magazine science fiction in the last twenty-five years. Still, undoubtedly, there will always be space operas, just as there will always be routine book and movie Westerns and whodunits.

The purely scientific story, like the pure chess-problem in detective fiction, is already declining, and will, I think, decline further, relatively, if not absolutely. There may be a few more readers for a highly scientific story (for a really good example like *Mission of Gravity*, there probably are) today than twenty-five years ago, but the audience for the less scientific story has increased enormously. A few old-timers may lament the fact, but science fiction, once the preserve of the scientifically minded, has been taken up by a host of readers (I am one of them) who have not the type of education necessary to give the really scientific story full appreciation. Last year a magazine was launched with the intention of giving science fiction back to the scientists, but it failed within a year.

A much more fertile field is that of the speculative science fiction which I discussed in the last chapter. It has been suggested by at least one rabid enthusiast that the estate of Jules Verne deserves a share of the royalties of certain inventors, because he showed them what to invent. I doubt this; people had certainly thought that it would be interesting to travel around the moon or twenty thousand leagues under the sea long before Verne suggested it. But there is a realm of possibility that is not yet scientific possibility where the imagination of writers can be of real service, both in showing us what we should like to bring into being or to avoid, the utopias and reverse utopias, and in opening our eyes to what may be an aspect of the truth. All knowledge is a branch of philosophy before it becomes (if it ever does) a branch of science; the atomic structure of the universe was a matter of philosophic theory for Lucretius; for us it is a matter for experimentation in the laboratory. Conditions on Venus are still to some extent a matter of speculation, but it seems safe to predict that within the foreseeable future they can be investigated at first hand. Apparitions of the dead—ghosts, if you like—communication at a distance, prediction of the future, all these were not long ago matters for speculation, if not for downright scoffing; now they are all coming under scientific examination. And if ever these phenomena are as clearly diagrammed as atoms, there will be some new frontier to take their place.

As it increasingly addresses itself to the general, that is the not especially scientific, public, science fiction will probably deal more with this borderland. I hope so, for it is here that it has the least possibility of harm and the greatest possibility of good: the least possibility of harm because in this admittedly uncertain region it cannot offer us the falsehoods that we have either everything to hope for or everything to fear from the abstraction called science; the greatest possibility of good because it is here, and not in suggesting a submarine, that one imagination may strike a spark from another. To illustrate what I mean I may cite a book that appeared before its time and has recently been reissued, William Sloane's *To Walk the Night*. In this book the excitement of a crowd at a tense football game becomes the power that brings into focus, so to speak, what becomes a supernormal or paranormal visitation. As Sloane uses it, that becomes much more than a device for the suspension of disbelief; it brings to mind Flammarion's statement that wherever there is a poltergeist there is an unhappy adolescent, and his suggestion that emotional disturbances can throw stones. Science fiction at its best releases the imagination by leading us to ask ourselves new questions—"Where does the physical end and the psychical begin?"—and by suggesting answers about which we feel that the truth, if it is ever discovered, will prove to be in some way *like* that. That is part of what William Blake meant when he said, "Everything that is possible to be believed is an image of the truth."

Here we are obviously getting close to the line between science fiction and fantasy. In measure as science fiction has moved away from the realm of pure science, it has moved toward that of fantasy. At least three-quarters of the material now published as science fiction is based on ideas that are not in the strictest sense scientific at all, but merely non-earthly, though there is usually some sort of scientific scaffolding about how these conditions arose or in what state the hero found them. This tendency, I believe, will increase. There will be a core of pure scientific fiction for those who like it, but more and more fantastic science fiction for those who like that; and, as to that, we have the word of Anthony Boucher, editor of *The Magazine of Fantasy and Science-Fiction*: "Most people prefer fantasy to science fiction, only they don't know it." That is, to be acceptable today, fantasy must have a pretense of science about it. People are quite willing to read about werewolves, but you must give them some sort of explanation about genes and chromosomes. Science fiction has in fact done for us what the translation of the *Arabian Nights* did for an eighteenth century that did not know how tired it was of the Age of Reason; it has brought imaginative writing back into repute. The fanciful imagination is like the Life Force at the end of *R.U.R.*; it may be kept under for a time, but in the end it will burst out somehow. For the Age of Reason, the channel was the *Arabian Nights*; as soon as people discovered that it was quite proper to read about jinn and rocs if they were in the Far East, they could not get enough of them; there was a flood of imitations, *Chinese Tales, Persian Tales, Hindoo Tales, Tales of the Genii,* stories whose oriental coloring was palpably artificial.

In the same way, our generation did not know how tired it was of twentieth-century realism until we discovered that it was respectable to read about marvels if they occurred in the future or on a distant planet. And we are willing to do with less and less in the way of scientific coloring; Ray Bradbury's robots and his Mars pay the merest lip service to science. Sometimes this invocation of the aegis of science is carried so far as to be funny. Recently there appeared a book called *The Fellowship of the Ring*, which, though intended for an adult audience, is a pure fairy tale; its characters are elves and enchanters (and not a hint of a gene or a chromosome among the lot) but even that carries on the jacket a quote saying, "This is really super-science-fiction." That is, of course, sheer nonsense; it is hard to know where to draw the line defining science fiction, but it certainly excludes *The Fellowship of the Ring*. What the critic actually means is, "This is imaginative writing, but you needn't be ashamed to be seen reading it." It is even possible that there may come a semantic shift, and that, as "comic books" bear no relation whatever to comedy, the term science fiction may come to be used indiscriminately for all types of nonrealistic fiction. But that is neither desirable nor likely. What is more likely is, as I have suggested, a clearer division into types. Actual science fiction, taken as a whole, has in it less science, just as the detective story has moved away from the pure chess problem; but in addition, as the detective story has given birth to the "novel of suspense" or murder without mystery, a somewhat different form and so labelled, so beside science fiction itself there is emerging a school of pure fantasy fiction.

It continues and will continue to owe some of its shape to science fiction. This fantasy fiction, and the science fiction that is fantasy in all but name, are already giving us the myths of our generation. A myth may be defined as a story which contains a truth, not in the manner of a fable, which illustrates a truth by anecdote, nor in the manner of an allegory, which translates a truth into other terms, but in the very nature of the story itself. In the great myths, like

Baldur and the Mistletoe or Kafka's *The Castle* or Stapledon's *Last and First Men,* no sort of translation is possible. You feel an inner meaning, but it cannot be expressed in other terms than in those of the story itself. And even in lesser myths, if you try to "give the meaning in your own words," as schoolmasters impossibly ask you to do with a poem—if you say, for instance, "The story of King Midas and the Golden Touch means that avarice is self-defeating," you will find that, as with the poem, much, indeed almost everything, has been lost. Myths today are being written in this form of imaginative writing. In discussing *R.U.R.* I tried to show how many myths it recalls or restates. The same thing is true with all the deepest science fiction. That Man is a creature with awesome potentialities for achievement and for self-destruction, and that the inhabitants of Earth are not the only powers in the universe—these are truths that men have never been able to forget for more than a generation or two. It is science fiction which is telling them to us now.

Suggested Reading

CHAPTER ONE

These suggestions, here and after the later chapters, make no pretense to completeness. They are merely samples, out of a very wide field, which I think will be of interest to anyone who wishes to go farther into the subject.

CRITICAL AND BIBLIOGRAPHICAL

Modern Science Fiction. Edited by Reginald Bretnor. Coward-McCann. A symposium.

Pilgrims through Space and Time. By J. O. Bailey. Argus Books. A history, chiefly of the predecessors of modern science fiction.

Science-Fiction Handbook. By L. Sprague de Camp. Hermitage. A book on writing science fiction.

The Checklist of Fantastic Literature. By Everett F. Bleiler. Shasta. A bibliography.

FICTION BEFORE THE MODERN PERIOD

Frankenstein. By Mary Shelley.

Twenty Thousand Leagues under the Sea. By Jules Verne.

Flatland, by A. Square. By Edwin A. Abbott. Dover Publications.

R.U.R. By Karel Capek.

Seven Science Fiction Novels. By H. G. Wells. Dover Publications.

Contains: *The Time Machine*
The Island of Doctor Moreau
The Invisible Man
The War of the Worlds
The First Men in the Moon
The Food of the Gods
In the Days of the Comet

CHAPTER TWO

This chapter has been devoted chiefly to a clearing of the ground, and very little of the material it covers is worth reading. The most reliable purveyor of space operas is E. E. Smith, who has many devotees among the fans. His main work falls into two series, as follows:

Triplanetary
Galactic Patrol
First Lensman
Gray Lensman

Second Stage Lensman
Children of the Lens

and

Skylark of Space
Skylark III
Skylark of Valeron

All these are published by Fantasy Press with the exception of *Skylark of Space*, which is published by the Buffalo Book Co.

CHAPTER THREE

The distinction between science fiction which is chiefly scientific and that which is more speculative is by no means a sharp one, and some of the books suggested for the later chapters might well have been included here. These, however, seem to me to illustrate the more scientific approach.

Astounding Science-Fiction Anthology. Edited by John W. Campbell. Simon & Schuster. A collection from *Astounding Science Fiction*, one of the best of the magazines, and the most scientific in its attitude.
Who Goes There? By John W. Campbell. Shasta.

Ice World. By Hal Clements. Gnome.
Needle. By Hal Clements. Doubleday.
Mission of Gravity. By Hal Clements. Doubleday.
City in the Sea. By Wilson Tucker. Rinehart.

CHAPTER FOUR

To the End of Time. By Olaf Stapledon. Funk and Wagnalls. An omnibus containing:
  Last and First Men
  Star Maker
  Odd John
  Sirius
  The Flames
I, Robot. By Isaac Asimov. Gnome.
The Demolished Man. By Alfred Bester. Signet.
Bring the Jubilee. By Ward Moore. Farrar, Strauss & Young.
Mutant. By Lewis Padgett. Gnome.
City. By Clifford Simak. Gnome.
Slan. By A. E. Van Vogt. Simon & Schuster.

CHAPTER FIVE

HUMOROUS

Lest Darkness Fall. By L. Sprague de Camp. Holt.

The Sinister Researches of C. P. Ransome. By H. Nearing.
  Doubleday.

SERIOUS

Children of Wonder. Edited by William Penn. Simon &
  Schuster. Contains *That Only a Mother.*

The Humanoids. By Jack Williamson. Simon & Schuster.
  Contains *With Folded Hands* and its sequel.

CHAPTER SIX

The distinction between fantasy and speculative science
fiction, like that between the latter and scientific science
fiction, is by no means sharp, and as with Chapter Three,
some of these titles might have been suggested for Chapter
Four. They seem to me to indicate some of the lines along
which science fiction is developing.

The Best of Fantasy and Science Fiction. Series I–IV.
  Edited by Anthony Boucher and J. Francis McComas.

An annual collection from one of the best of the magazines.

Worlds of Wonder. Edited by Fletcher Pratt. Twayne. An anthology.

Stories for Tomorrow. Edited by William Sloane. Funk & Wagnalls. An anthology.

The Martian Chronicles. By Ray Bradbury. Doubleday.

Childhood's End. By Arthur C. Clarke. Ballantyne.

The Incomplete Enchanter. By L. Sprague de Camp and Fletcher Pratt. Holt.

To Walk the Night. By William Sloane. Dodd Mead.